my guide to earth's habitats

Ocean

Susan H. Gray

CHERRY LAKE PRESS

Published in the United States of America by Cherry Lake Publishing Group
Ann Arbor, Michigan
www.cherrylakepublishing.com

Reading Adviser: Beth Walker Gambro, MS, Ed., Reading Consultant, Yorkville, IL
Book Design: Jennifer Wahi
Illustrator: Jeff Bane

Photo Credits: © Zaie/Shuttershock.com 2, 3, 24; © Vlad61/Shuttershock.com, 5; © Rich Carey/Shuttershock.com, 7; © Johan Holmdahl, 9; © Choksawatdikorn/Shuttershock.com, 11; © Imagine Earth Photography/Shuttershock.com, 13; © Isabelle Kuehn/Shuttershock.com, 15; © Martin Prochazkacz/Shuttershock.com, 17; © Pong471/Shuttershock.com, 19; © superjoseph/Shuttershock.com, 21; © Enessa Varnaeva/Shuttershock.com, 23; Cover, 6, 14, 22, Jeff Bane

Cherry Lake Press is an imprint of Cherry Lake Publishing Group.

Library of Congress Cataloging-in-Publication Data

Names: Gray, Susan Heinrichs, author. | Bane, Jeff, 1957- illustrator.
Title: Ocean / Susan H. Gray ; illustrated by Jeff Bane.
Description: Ann Arbor, Michigan : Cherry Lake Publishing, 2022. | Series: My guide to earth's habitats | Audience: Grades K-1
Identifiers: LCCN 2022005384 | ISBN 9781668910597 (paperback) | ISBN 9781668908990 (hardcover) | ISBN 9781668913772 (pdf) | ISBN 9781668912188 (ebook)
Subjects: LCSH: Marine ecology--Juvenile literature. | Ocean--Juvenile literature.
Classification: LCC QH541.5.S3 G74 2022 | DDC 577.7--dc23/eng/20220214
LC record available at https://lccn.loc.gov/2022005384

Printed in the United States of America
Corporate Graphics

table of contents

About the author: Susan H. Gray has a master's degree in zoology. She loves writing science books, especially about animals. Susan lives in Arkansas with her husband, Michael. She thinks the oceans are fascinating places.

About the illustrator: Jeff Bane and his two business partners own a studio along the American River in Folsom, California, home of the 1849 Gold Rush. When Jeff's not sketching or illustrating for clients, he's either swimming or kayaking in the river to relax.

Full of Life

Oceans cover most of Earth. Every ocean is home to living things.

The sun warms the ocean's **surface**. Deeper down, things change. The water is dark and cold.

Algae live in the sunlight near the surface. Some grow into long **strands**.

Tiny animals also drift there. They make up **plankton**.

Fish eat plankton. So do blue whales.

Corals grow on the ocean floor. Together, they form a **reef**. Sea turtles feed at the reef.

Some animals like the open water. Whales swim freely here. **Dolphins** do, too.

Deep down, it is totally dark. Fish cannot see to hunt. So some make their own light.

These fish have a special **organ**. It lights up. Small fish swim toward it. The big fish then snap them up!

Oceans are amazing. They are full of life. Everything has just what it needs.

glossary & index

glossary

algae (AL-jee) plantlike organisms that include seaweeds

corals (KOR-uhlz) hard skeletons or cups made by soft-bodied ocean animals

dolphins (DOL-fuhnz) ocean mammals that breathe air and have fishlike bodies

organ (OR-guhn) a body part that has a particular task

plankton (PLANGK-tuhn) small aquatic plants and animals that drift in the water

reef (REEF) a ridge of rock or coral

strands (STRANDZ) long, stringy fibers

surface (SUR-fuhs) the top layer of something

index